The Wonder of It All

This book is for Charlie, my newborn son

One month old, his journey's just begun

It's also for James, for Maya and Sergio

May you have wonder wherever you go

"Star Child" by Filipov Ivo. Credit: Filipov Ivo

Who are you? What makes you, you?
I'd love to know, please give me a clue!
You might not know it, but you're really quite unique
Like the universe we live in and the world at your feet

Elementary school children in Paris, France. Credit: R. K. Singam

But I know what you're thinking, there's others here like you
Same hair, same eyes - why, there's a hundred just at school
That may seem true but look again, you're different
Different hopes, different dreams and that's really something brilliant

The first image of the whole Earth taken by humans. Credit: NASA

Where do you live? What town or what city?
Do the stars shine? If not, that's a pity
How big is your home? How many live there too?
Did you know our planet is unique like you?

Brown Park, Louisville Kentucky. Credit: Richard J. Bartlett

It's called the Earth and it's our place in space
It's where we all live, the whole human race
A medium-sized planet, the only one with life
We have rivers and forests and many things you like

Our Sun, as photographed by NASA's Solar Dynamics Observatory. Credit: NASA/SDO (AIA)

What of your parents? Do you have both or just one?

The Earth has no mother but it does have the Sun

The Sun is a star, it keeps us alive

Without it we're nothing, we couldn't survive

The waxing crescent Moon. Credit: Richard J. Bartlett

And after the Sun sets, what brightens our skies?
What light can you see with only your eyes?
It changes quite quickly, from a crescent through full
Then back to a crescent before vanishing from view

The inner planets to scale. Credit: Scooter20, from NASA images. (Source, Wikimedia)

Do you have any siblings? Any sisters or brothers?
The Earth is one world but there are seven others
First is Mercury, then it's Venus, Earth and then Mars
It seems like a lot but we're still far from the stars

The outer planets to scale. Credit: Lunar and Planetary Institute

After these planets we journey out again
By Jupiter! Then Saturn! These giant worlds reign!
But wait, we're not there yet! There's two more to go
Uranus then Neptune, we're a long way from home

Mercury, as photographed by the MESSENGER space probe. Credit: NASA (color by Jason Harwell.)

How fast can you run? Are you slow? Are you quick?
Mercury is fastest, it's really quite slick
It's the first of the planets and it's nearest the Sun
If you're smart you can see it, it's really good fun

An artist's impression of the surface of Mercury. Credit: Michael Carroll/The Johns Hopkins University Press, 2008.

Sometimes it's there in the evening twilight
It appears for a few days then disappears from sight
Wait a short while and it'll show up again
It shimmers in the morning, the games never end!

Venus, as photographed by the Pioneer Venus space probe. Credit: NASA/NSSDC

Do you like warmth? Do you bathe at the beach?
There's a planet that's hot, it has plenty of heat
Next from the Sun, Venus is nearest the Earth
But there's no ocean here, there's nothing but dirt

The crescent Moon and Venus in the evening sky. Credit: Belojica Mile

It's not even sunny, you won't see the stars
The skies are all cloudy, not clear blue like ours
It isn't much fun but it's pretty to spy
When it's close to the Moon in the evening sky

Mars, photographed by the Hubble Space Telescope. Credit: NASA/ESA/The Hubble Heritage Team (STScI/AURA)

Have you ever lost something that can't be replaced?
Something so precious it made your heart ache?
Mars once had water, or so we believe
It's now cold and dry and for that we should grieve

The surface of Mars photographed by the Spirit rover. Credit: NASA/JPL-Caltech/Cornell University

For a long time ago, on a planet that's red
Life may have begun but now it's long dead
But what if it hadn't - there'd be people there too
Cousins looking back and waving down at you!

A montage of Jupiter and its four largest moons, imaged by the Voyager 2 space probe. Credit: NASA/JPL

How big are you? Are you short? Are you tall?
In our system of planets, one world dwarfs them all
Jupiter is huge, it's really impressive
But it nurtures a storm you don't want to mess with

The Great Red Spot on Jupiter, imaged by Hubble. Credit: M. Wong and I. de Pater (University of California, Berkeley)

That Red Spot you see, it's larger than Earth!
It's been there for years - long before your birth
You might think it's windy on a bad stormy day
But the winds there on Jupiter could blow you away!

A montage of Saturn and its moons, imaged by the Voyager 2 space probe. Credit: NASA/JPL

What looks amazing? What makes you gasp "wow!"?

Maybe a sunset or a white fluffy cloud?

Try seeing Saturn on a clear starry night

Through a telescope it's astounding, it's really a sight

An artist's impression of the icy chunks that form the rings of Saturn. Credit: Anynobody (Source: Wikimedia)

Saturn has rings, but you probably know that
They're made of ice chunks, but from Earth they look flat
Millions of snowballs all orbiting the planet
Some of them are bigger than the house you inhabit!

A montage of Uranus and its moons, as imaged by the Voyager 2 space probe. Credit: NASA/JPL

Are you an oddball? Do you do crazy things?
Do you have fun? Do you laugh? Do you sing?
Do you roll around in the grass or the snow?
There's a planet like you - it's odd - did you know?

An artist's depiction of Uranus rising over the surface of its moon Miranda. Credit: NASA/JPL

Uranus is seventh in line from the Sun
But it rolls on its side, it's a curious one
It doesn't sit upright like most of its kind
It does its own thing and that's perfectly fine

Neptune, as imaged by the Voyager 2 space probe. Credit NASA/JPL

What calms you down? What brings you peace?
Is it your mother? Your father? Or a kiss on the cheek?
Or maybe a color? Like cool, icy blue
Take a look at this world; it's name is Neptune

An artist's depiction of Neptune and its largest moon Triton. Credit: Jcpag2012 (From Wikimedia)

It's the last of the planets and we're a long way from home
Out here it's peaceful, where few others roam
But Neptune isn't lonely, even though it's quite far
The Earth is very distant, the Sun a faint star

Pluto, imaged by the New Horizons space probe. Credit: NASA/John Hopkins University/Southwest Research Institute

What of the cold? The snow and the ice?

Do you shiver like a snowman or do you think it's quite nice?

If so, consider Pluto, it might be the one

A small rock that's frozen with no warmth from the Sun

An artist's depiction of Pluto's largest moon, Charon, rising over the horizon. Credit: ESO/L. Calçada

In the darkness of space, this world orbits slowly
It once was a planet but now it's quite lowly
Pluto was different from the other worlds we know
It was sadly demoted a few years ago

The constellation Ursa Major, with the Big Dipper visible near the top of the image. Credit: Roberto Mura

Who are your friends? I bet you have many
They stand by your side when your heart becomes heavy
If you look to the stars, they're there for you too
Sometimes it's cloudy, but they're shining on you

The constellation Orion. Credit: Roberto Mura

Through winter and summer, the fall and the spring
For thousands of years they've seen everything
Rising and shining through our triumphs and defeats
Life comes and goes but the stars never leave

The red dwarf star Proxima Centauri with Alpha Centauri in the background. Credit: Rept0n1x (From Wikimedia)

Do you have neighbors? Do you visit next door?
The Sun is one star but nearby are more
Which one is closest? That's Proxima Centauri
A bit of a mouthful but that's a long story

Sirius, with its blue white-companion in the background. Credit: NASA, ESA and G. Bacon (STScI)

Have you heard of Sirius? Brightest in the sky
The Egyptians seemed to like it and it's also nearby
It glitters like a diamond through the cold depths of winter
It brightens our nights 'til the weather is warmer

The Orion Nebula. Credit: NASA/ESA, M. Robberto (STScI/ESA), Hubble Space Telescope Orion Treasury Project

When were you born? Were you born at home?
Are you a twin? Stars are never born alone
They form from gas clouds, like a nursery in space
These stars are like babies in a mother's embrace

The Pleiades star cluster in Taurus. Credit: Boris Stromar

Things start to change as the siblings grow older
Some drift apart while some stick together
These clusters of stars glint like gem stones on satin
Some burn white but all glow with passion

An artist's depiction of a fictitious Earth-like ocean planet. Credit: Luciano Mendez

And when you grow older, where will you be?
Will you stay near home? Have your own family?
Some stars have planets, they're almost like children
But we haven't found life, that's still yet to happen

An artist's depiction of the red giant star Betelgeuse in Orion. Credit: ESO/L. Calçada

You think stars live forever, that stars always shine
But nothing's immortal and stars sometimes die
Some swell to red giants, then shrink down again
But others die differently, they have another end

An artist's depiction of a supernova. Credit: NASA

What makes you angry? What really makes you mad?
Do you ever explode and try to be bad?
Some stars blow up but it's not due to a tantrum
It happens when they're old and it's not really random

The Crab Nebula - the remains of supernova observed in 1054 AD. Credit: NGC 1535 (From Wikimedia)

It happened before, a thousand years ago
A nova appeared and through daylight it glowed
We saw the remains a few centuries later
We still see it today, it's called the Crab Nebula

An artist's depiction of a black hole consuming a star. Credit: ESA/NASA/Felix Mirabel

What do you eat? Maybe sweet candy bars?
Did you know there's a thing that swallows whole stars?
It gobbles up everything, it simply never stops
If you did that you'd regret it, if you did that you'd pop!

An artist's depiction of a black hole close-up. Credit: Dmytro Ivashchenko

It's called a black hole and its name is quite right
For nothing escapes it, no, not even light
I know it sounds dangerous, but trust me, don't worry
They're nowhere near us, they're all light years away

The summer constellations with the misty grey Milky Way. Credit: Steve Jurvestson

What country do you live in? Are there many towns and cities?
The Sun is one of billions in our Milky Way galaxy
A faint misty river, you can see it in the summer
Wait a few months and you might see another

The Andromeda Galaxy is the closest major galaxy to our own. Credit: Rodrigo Ferraz Castiñeiras

This galaxy is found under cool autumn skies
It's the furthest thing seen with only your eyes
A sister in Andromeda, it's larger than our own
With billions more stars and maybe an alien home

These two galaxies are slowly colliding. Credit: NASA, ESA, Hubble Heritage (STScI/AURA)-ESA and K. Noll

How far have you travelled? What things have you seen?
Galaxies are many, but what does this mean?
You may feel small, but please try to remember
You are unique, there's really no other

"Star Child" by Filipov Ivo. Credit: Filipov Ivo

Your place is here, with one life to live
Make it a good one, give all you can give
Change your world, answer life's great call
You have the power - that's the wonder of it all

Appendix

Why I Wrote This Book

Following the publication of my last book, *Easy Things to See With a Small Telescope* on December 1st, 2015, I started work on my next astronomy book. It wasn't a children's book but instead a more mainstream piece detailing deep sky objects. I'd completed a lot of research already and only had to complete some observations of autumnal constellations. Life, of course, had other plans.

Firstly, my telescope was broken in an accident, which quickly put the brakes on my book. At least until next autumn. Then, my second son Charlie was born, about three weeks early, on December 18th. We celebrated Christmas and New Year and, not wanting to sit still and be writing *nothing* I decided I needed to write *something.*

You don't get much time to write anything with a newborn in the house. I needed to write something that could be short and that I could work on randomly, as the opportunities presented themselves. (Thanks to Google Docs and my cell phone!) I thought of Charlie, I wondered where he would go, what he would see, who he'd become… and to some extent this book came about by itself.

Notes for Parents & Additional Information

N.B. If you're interested in tracking the Moon and planets, please refer to the latest edition of my book, *The Night Sky Sights*, which I publish annually. The book details astronomical events you can see throughout the year with only your eyes and highlights the Moon and visible planets (Mercury through to Saturn), meteor showers and constellations.

If you'd like to see a sample PDF of the current or upcoming month, please feel free to email me at astronomywriter@gmail.com

If you have a small telescope and are looking for more objects to see in the night sky, please refer to my book, *Easy Things to See With a Small Telescope*. Again, please feel free to email me if you'd like to see a PDF sample.

Pages 6 & 7 - Earth

Earth is the only place we know that can support life and if an environmental disaster were to occur, there really would be nowhere else to go. As much as I didn't want the book to be negative, I thought it was important to at least hint at this and emphasize its uniqueness. (Incidentally, the image of the Earth is an adjusted version of the very first photo taken of the whole Earth. It was photographed by the Apollo 8 astronauts as they flew on the first manned mission to the Moon, December 1968.)

Page 8 - The Sun

The Sun is a pretty average star and is thought to be around 4½ billion years old. It's probably middle-aged, so it's got another 4-5 billion years to go. In comparison, modern humans have only been around for about 200,000 years and, one way or another, we'll be long gone before the Sun comes to the end of its life. Some might compare the Sun to a parent, but technically the planets weren't born from the Sun, they only formed within the same cloud of gas and dust. Since the central mass was largest, that became the Sun and everything else orbited it as a result.

Page 9 - The Moon
The Moon, of course, is easily visible throughout almost the entire month, either in the evening or morning sky. After new Moon, when the Moon cannot be seen at all, it begins to re-appear in the evening sky before reaching its full phase about two weeks later. At that time, it rises at sunset and sets at sunrise. After full Moon, it rises after sunset and won't set until the morning hours. Depending on how close it is to new Moon, it may be visible throughout the entire morning. It then turns new (again, about two weeks after full Moon) and then the cycle begins again.

Pages 10 & 11 - The Planets
Since the demotion of Pluto in 2006, there are officially now just eight planets with five of them (Mercury, Venus, Mars, Jupiter and Saturn) being visible with just your eyes. I didn't choose to write about the asteroids, comets and meteors because I felt the book was long enough as it was!

Pages 12 & 13 - Mercury
Mercury can be glimpsed in the evening and pre-dawn sky but it's not an easy planet to spot as it's always quite close to the horizon. Many folks go their entire lives never seeing the planet at all or, if they do, mistaking it for a moderately bright star.

Because the planet doesn't shine brilliantly, I didn't feature an image of it in the sky as it would have been potentially difficult to reproduce the image with the planet being easily identifiable. Consequently, I've provided an artist's rendition of the view from the surface, which is similar to the images provided for some of the other planets.

Pages 14 & 15 - Venus
Venus, on the other hand, is easily visible as a brilliant white star in evening or pre-dawn sky and the second image nicely illustrates this. Despite its apparent beauty, Venus is a hellish planet, with temperatures in excess of 750F (400C) and an atmospheric pressure great enough to crush almost anything that lands on its surface.

Pages 16 & 17 - Mars
To all intents and purposes, Mars is a dead world. In September 2015, NASA announced the discovery of flowing salt water but it's thought that water was much more abundant in the planets' distant past, when the Sun was hotter and the surface of Mars was warmer. It's possible that microbial life might have developed and might still exist but there's been no conclusive evidence as yet.

Pages 18 & 19 - Jupiter
Jupiter is, by far, the largest of the planets and could contain all the other planets within it. It has four large moons (as depicted) which can easily be seen and tracked with binoculars or a small telescope. If you have a telescope, you can also see the dark bands across its atmosphere. Unfortunately, you probably won't be able to see the Great Red Spot unless you're willing to invest in a slightly larger 'scope. (You'll also need to know when it will be facing us, but there's plenty of software that can help with that.)

Pages 20 & 21 - Saturn
Saturn is sure to generate gasps anytime someone sees it for the first time. It really is quite stunning when seen through a telescope and never disappoints. No one truly knows how the rings were formed, but they may be the shattered remains of an icy moon that strayed too close to the planet and was broken apart

by gravity. In addition, it has over 60 moons, many of which have been visited and imaged by the Cassini space probe.

Pages 22 & 23 - Uranus
Uranus is on the very edge of naked eye visibility, but you need keen eyesight and you have to know exactly where to look. Through a telescope it presents a tiny aquamarine disc. It really is a little odd as it's the only planet that has an axis that's tipped over by nearly ninety degrees. So it literally rolls around the Sun, with the north pole facing us for about half of its 84 year orbit and the southern pole facing us for the remainder of the time.

Pages 24 & 25 - Neptune
Neptune is now, officially, the last of the planets - although speculation continues to circulate that a ninth planet lies much further out. It has a moon, Triton (depicted in the second image) that actually has liquid nitrogen geysers in its surface. The nitrogen freezes the moment it comes into contact with the extremely thin atmosphere and then falls back down as snow.

Pages 26 & 27 - Pluto
Pluto, of course, was once classified as the ninth planet but it was always the odd one out. It was a tiny, rocky world beyond the four large gas giant worlds of Jupiter, Saturn, Uranus and Neptune. But apart from this, it had an odd orbit, at an odd angle that took the world closer to the Sun than Neptune. It also had a moon, Charon, which was about half the size of Pluto itself - a uniqueness that almost made it a double planet system. It just didn't fit in with the others.

So when astronomers discovered other similar worlds with similar characteristics, a controversial debate ensued: should all these worlds be classed as planets (and have a solar system with potentially hundreds of planets) or should Pluto be demoted into a new category of dwarf planets instead? In 2006 the astronomical community argued and voted (several times) and the rest, as they say, is history.

Page 28 - Ursa Major
The seven brightest stars in the constellation of Ursa Major (seen in the top of the image on this page) are famously known across almost the entire northern hemisphere. For example, in the United Kingdom, where they never set, they're known as the Plough whereas in North America observers know them as the Big Dipper.

Also well known is that you can use two of the stars to locate Polaris, the pole star, in the neighboring constellation of Ursa Minor. However, what many don't realize is that you can use the constellation to find many other nearby constellations.

You can find a star chart for Ursa Major at the end of this book.

Page 29 - Orion
Orion, another very famous constellation, dominates the winter sky and remains visible through at least the first half of spring. Like the seven brightest stars of Ursa Major, it can be used to locate nearby constellations. In particular, follow the three stars of the belt down to find Sirius, (see page 31 and below) the brightest star in the sky, in the constellation of Canis Major.

You can find a star chart for Orion at the end of this book.

Page 30 - Proxima Centauri
Proxima is part of the Alpha Centauri multiple star system and is slightly closer to us than the system's main pair of stars – hence the name *Proxima*. A red dwarf star, it lies at a distance of 4 ¼ light years and is too faint to be seen with just your eyes. However, the combined light of the Alpha Centauri system appears as the third brightest star in the night sky and can easily be seen from the southern hemisphere from May to August.

Page 31 – Sirius
Sirius, the brightest star in the sky, is prominently visible throughout the winter and early spring. It's easily found by following the three stars of Orion's belt downward toward Canis Major, the constellation in which it resides. (See the back of the book for a chart depicting Orion.) Canis Major is also known as the Great Dog and is one of Orion's hunting dogs which, in turn, gives Sirius its nickname as the Dog Star.

The Egyptians associated the star with the goddess Isis and its early morning rising coincided with the flooding of the Nile. It's a hot white star, only eight light years away, with a small, white dwarf companion (nicknamed "the pup.")

Page 32 - The Orion Nebula
The Orion nebula marks the sword of Orion and is easily visible under the three stars of Orion's belt as a tiny, misty patch. Even observers in the suburbs can glimpse it with just their eyes. This is probably the most famous example of a nebula, a huge cloud of gas and dust in space, some 1,300 light years away and some 20 light years across. It's from these clouds that stars are born.

In the center of the nebula is a tiny cluster of brilliant, hot young stars known as the Trapezium. You can see two of these stars with binoculars while a small telescope at low power should reveal several more. These are stars born from the nebula itself and, as time progresses, these stars will leave the nebula behind and venture out into space. The Sun itself was born from a nebula like this and some believe the Orion Nebula might have even been its birthplace.

(If you have a telescope, take some time to observe the Orion Nebula. You won't see the fantastic colors that apparent in photographs, but it remains simply stunning and not to be missed!)

Page 33 - The Pleiades
Many star clusters are relatively young as the individual stars were born together from the same nebula. Sometimes the stars will drift apart while other clusters will remain gravitationally bound to one another. The Pleiades is probably the best and most famous example of one such cluster. Only about 100 million years old (compared to our 4 ½ billion year old Sun), the Pleiades have been known since antiquity and are easily seen from autumn through to early spring.

Like Sirius and the Orion Nebula, you can use the three stars of Orion's belt to find them. This time, draw an imaginary line upwards to the orange star Aldebaran, which marks the eye of Taurus the Bull. (Incidentally, Aldebaran marks the edge of another cluster, the V shaped Hyades – easily seen with the eye but great in binoculars.) Keep going until you come to a tiny grouping of five or six stars. These are the Pleiades. Again, like the Hyades, they're easily seen with just your eyes but are best observed with binoculars.

In the image you can see blue misty patches surrounding the stars. This was originally thought to be the remnants of the nebula from which the stars were born but it's now thought the stars are simply moving through the cloud as they journey through space. Unfortunately, the nebula can be very tricky to see unless you're under very clear, dark skies and typically only shows up well in photographs.

Page 34 – Exoplanets

To date, over 2,000 planets have been discovered orbiting some 1,300 stars. Known as *exo*planets because they lie outside our own solar system, none of these worlds are believed to be Earth-like although there are a handful that come relatively close. Most are huge gas giants, like Jupiter and Saturn, with the vast majority existing outside their parent star's habitable zone.

Of the planets known, one of the most likely candidates for being Earth-like is Kepler-62 f. At just under 1 ½ times the size of the Earth, it orbits a red dwarf star some 1,200 light years away. It has a "year" of 267 days and may have an ocean that covers the entire planet. That possibilities for life on this world are intriguing, to say the least!

Page 35 - Red Giants

Towards the end of their lives, many stars will swell up to a size much larger than normal and will simultaneously start to cool. As their surface temperatures drop, so their color reddens, in much the same way that white-hot metal dulls to yellow, orange and then red as it cools.

The image on this page depicts Betelgeuse, the orange-red star that marks the right shoulder of Orion, the hunter, in the winter sky. (See page 29 and the star chart at the back of this book.) Betelgeuse is one of the largest stars known. If it were placed at the center of the solar system, it would easily extend past the asteroids and could swallow the planets Mercury, Venus, Earth and Mars within it. Likewise, once the Sun reaches the end of its life (in a little over five billion years) it too will become a red giant and will may well expand to swallow the Earth.

Depending on the mass of the giant, it will either then shrink back down to a white dwarf star (the Sun's likely fate) or may well explode as a nova.

Page 36 – Novae

A nova occurs when a giant star, such as Betelgeuse, reaches the end of its life. The core is no longer able to prevent itself from collapsing under its own gravity and the resulting explosion causes the dying star to suddenly appear thousands of times brighter than normal. Supernovae occur when the star is abnormally massive and the resulting explosion may be visible even in daylight. The new star (nova) will often be seen for weeks, sometimes months, before slowly fading and disappearing from view. Naked eye supernovae are especially rare and only three have been observed over the past thousand years.

Page 37 - The Crab Nebula

The remains of a supernova explosion that was observed in July 1054, the Crab Nebula is a favorite with amateur astronomers and is one of the few example of its kind that can be relatively easily observed in the night sky. It's visible to almost anyone with a small telescope but you'll need to be familiar with the constellations and away from the lights of any nearby towns and cities.

First discovered in 1731, it wasn't linked to the supernova until the early 20th century. Observations showed the nebula to be expanding and its position was found to match that of the supernova. It's thought to be about 6,500 light years away and about 13 light years in diameter.

Page 38 - 39 Black Holes
There are several possible outcomes following the explosive self-destruction of a star in a supernova. Depending upon the original mass of the star, the core might collapse and form a neutron star or – in the case of a particularly massive star – the core might collapse even further and form a black hole. The reason it's black is because light is a form of electromagnetic radiation and the particles are unable to travel fast enough to escape the black hole's immense gravitational pull. Black holes can, theoretically, consume almost anything they come into contact with, including any stars, planets and moons that might be nearby. Fortunately for us, the nearest black hole is thought to be some 1,600 light years away and we're well beyond its gravitational pull.

Page 40 - The Milky Way
The Milky Way is our own, home galaxy and, theoretically, can be seen at any time of year as a faint misty grey river of stars that spans the sky from one horizon to another. It's best seen in the summer because the heart of the galaxy is located close to Sagittarius, a summer constellation. You may be able to pick out the teapot-shaped group of stars just above the horizon in the middle of the image, with the heart of the galaxy appearing as a brighter patch like steam from the teapot's spout. If you want to see this for yourself, make sure you're far away from the lights of any nearby towns and cities!

Page 41 - The Andromeda Galaxy
The Andromeda Galaxy is the most distant object you can see with just your eyes. You can see it – if you know where to look – under dark, autumn skies but you really need to be away from any nearby bright lights. If you live in the suburbs of a town or city, you probably won't see it without optical aid but it should easily be visible with binoculars or a small telescope. Regardless of how you observe it, it typically appears as an elongated, faint grey smudge against the sky. Because of its faintness, it really requires beginners to familiarize themselves with the constellations of autumn first.

Page 42 - The Universe
No one knows how many galaxies are in the universe. If the universe is truly infinite, then logically the number of galaxies, stars, planets and even life forms must be infinite also. There are, of course, a lot of arguments - scientific, philosophical and religious - regarding the nature of the universe and I will not take the time to debate them here. I will merely state that I believe the universe to be far stranger than we could possibly imagine!

About the Images
With the exception of the image of Brown Park in Louisville, the crescent Moon and my author photo (all of which are my own) all the images in this book are listed on Wikimedia as being freely available for use by others, provided that proper credit is given. In the vast majority of cases, I've applied an effect to the images to give them the appearance of a painting. The only exception is the *Star Child* image as this is already a painting and my author photo.

I make no claims to ownership of any of the images, except for those I specify in the paragraph above. If you own the copyright on any image and you'd like it removed from the book, please feel free to contact me. Likewise, if the image accreditation is incorrect, please let me know. My email address is astronomywriter@gmail.com

About the Author

Photo by James Bartlett

I've had an interest in astronomy since I was six and although my interest has waxed and waned like the Moon, I've always felt compelled to stop and stare at the stars.

In the late 90's, I discovered the booming frontier of the internet, and like a settler in the Midwest, I quickly staked my claim on it. I started to build a (now-defunct) website called *StarLore*. It was designed to be an online resource for amateur astronomers who wanted to know more about the constellations - and all the stars and deep sky objects to be found within them. It was quite an undertaking.

After the website was featured in the February 2001 edition of *Sky & Telescope* magazine, I began reviewing astronomical websites and software for their rival, *Astronomy*. This was something of a dream come true; I'd been reading the magazine since I was a kid and now my name was regularly appearing in it.

Unfortunately, a financial downturn forced my monthly column to be cut after a few years but I'll always be grateful for the chance to write for the world's best-selling astronomy magazine.

I emigrated from England to the United States in 2004 and spent three years under relatively clear, dark skies in Oklahoma. I then relocated to Kentucky in 2008 and then California in 2013. I now live in the suburbs of Los Angeles; not the most ideal location for astronomy, but there are still a number of naked eye events that are easily visible on any given night.

Also by the Author…

2016 An Astronomical Year is written for everyone with an interest in astronomy and contains information on hundreds of night sky events throughout the year. It was designed for astronomers of all levels and includes details of the lunar phases and eclipses, as well as conjunctions, oppositions, magnitude and apparent diameter changes for the planets and major asteroids.

To date, the 2015 edition has been downloaded nearly 3,000 times, was ranked #1 in Free Kindle Astronomy books, within the Top 10 Paid Kindle Astronomy books and within the Top 50 Free Kindle Non-Fiction books.

It is available in paperback and Kindle editions in the United States, Canada and the United Kingdom. (Please be aware that due to the cost of printing in color, the paperback does not contain images and is purely text only.)

2016 The Night Sky Sights is specifically designed for absolute beginners and casual stargazers without a telescope. The guide highlights over 125 astronomical events in 2016 - all of them visible with just your eyes - and showcases events visible in both the evening and pre-dawn sky as well as those you can see throughout the night.

It is currently available in paperback and Kindle editions in the United States, Canada and the United Kingdom.

The Astronomical Almanac (2016-2020): A Comprehensive Guide to Night Sky Events provides details of thousands of astronomical events from 2016 to the end of 2020. Designed for more experience astronomers, this the guide includes almost daily data and information on the Moon and planets, as well as Pluto, Ceres, Pallas, Juno and Vesta.

To date, the 2015-2019 edition has been downloaded nearly 6,000 times, was ranked #1 in the Free Kindle Astronomy book category, #3 in the Paid Kindle Astronomy book category and within the Top 50 of *all* Free Kindle books in October 2014.

It is available in paperback and Kindle editions worldwide, including the United States, Canada, the United Kingdom and Australia.

The Amateur Astronomer's Notebook: A Journal for Recording and Sketching Astronomical Observations is the perfect way to log your observations of the Moon, stars, planets and deep sky objects. It is available as both a full-size 8.5" by 11" journal and also as a 5" by 8" pocket notebook. The larger edition has room for 150 observing sessions while the pocket edition allows you to record 100 observations.

It is available as a paperback in selected areas. (Full Size Edition: United States, Canada and the United Kingdom. Pocket Edition: United States, Canada and the United Kingdom.)

The Deep Sky Observer's Guide offers you the night sky at your fingertips. As an amateur astronomer, you want to know what's up tonight and you don't always have the time to plan ahead. Maybe the clouds have suddenly parted. Maybe you're at a star party. Maybe you want to challenge yourself with something new but don't know where to start.

The guide can solve these problems in a conveniently sized paperback that easily fits in your back pocket. Take it outside and let the guide suggest any one of over 1,300 deep sky objects, all visible with a small telescope and many accessible via binoculars.

Easy Things to See With a Small Telescope: A Beginner's Guide to Over 60 Easy-to-Find Night Sky Sights – the #1 best-selling telescope book in the UK Amazon store, January 2016, it is specifically written with the beginner in mind and highlights stunning multiple stars, star clusters, nebulae and the Andromeda Galaxy.

Each object has its own page which includes a map, a view of the area through your finderscope and a depiction of the object through the eyepiece. There's also a realistic description of every object based upon the author's own notes written over years of observations. Additionally, there are useful tips and tricks designed to make your start in astronomy easier and pages to record your observations.

If you're new to astronomy and own a small telescope, this book is an invaluable introduction to the night sky.

It is currently available in paperback and Kindle editions in the United States, Canada and the United Kingdom.

Echoes of Earth – a collection of science fiction, mythological and philosophical short stories that I wrote many, many moons ago. (i.e., in the mid 1990's.)

It is available as a Kindle edition in selected areas. (United States, Canada, the United Kingdom and Australia.)

The Author Online

Email: astronomywriter@gmail.com

Amazon US: http://tinyurl.com/rjbamazon-us

Amazon UK: http://tinyurl.com/rjbamazon-uk

The Astronomical Year: http://tinyurl.com/theastroyear

Facebook: http://tinyurl.com/rjbfacebook

Twitter (@astronomywriter): http://tinyurl.com/rjbtwitter

Clear skies,

Richard J. Bartlett

January 31st, 2016

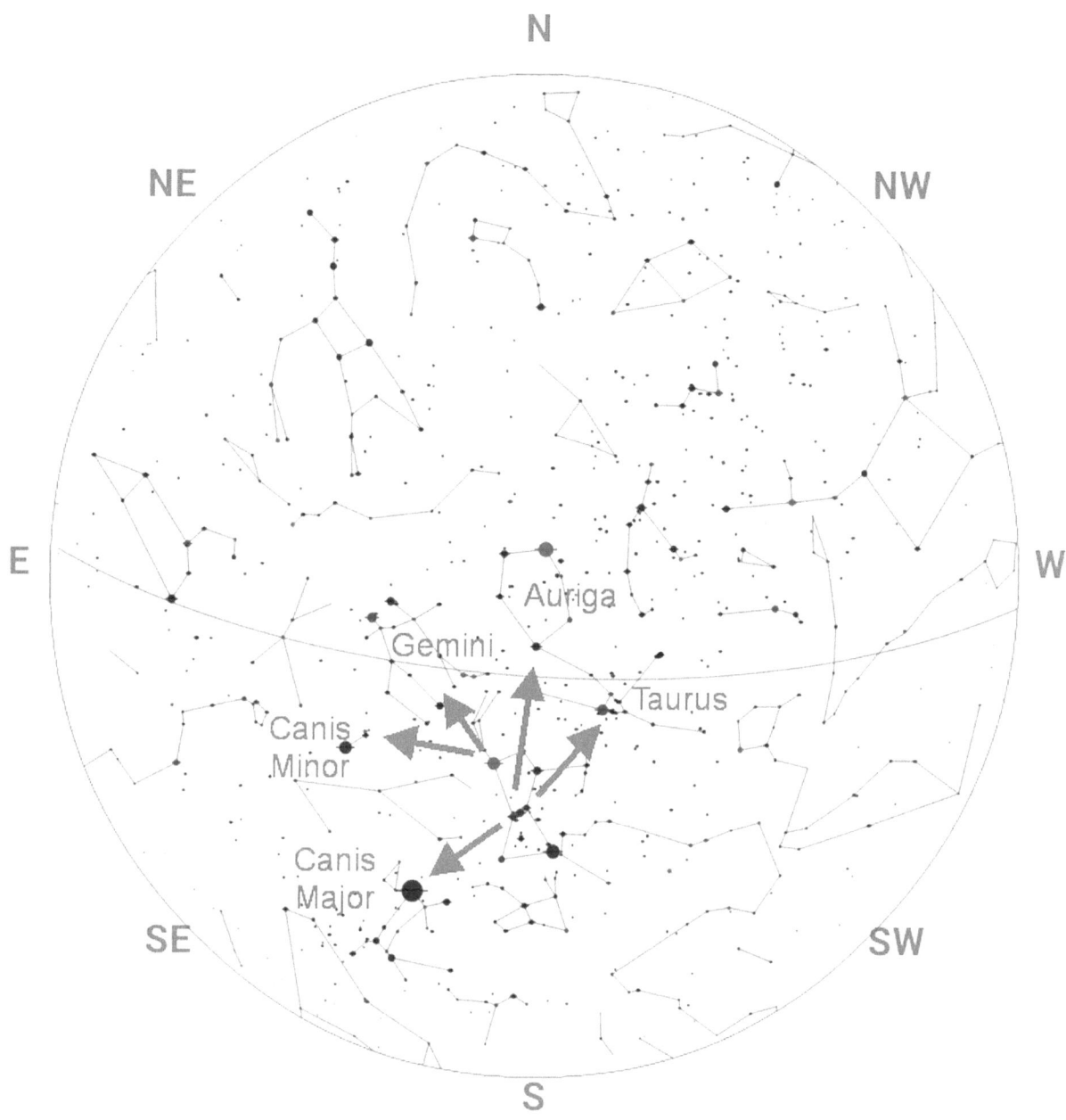

This chart can help you to use Orion to identify other constellations. It depicts the night sky at the following times:

Late January	Early February	Late February	Early March
10 p.m.	9 p.m.	8 p.m.	7 p.m.

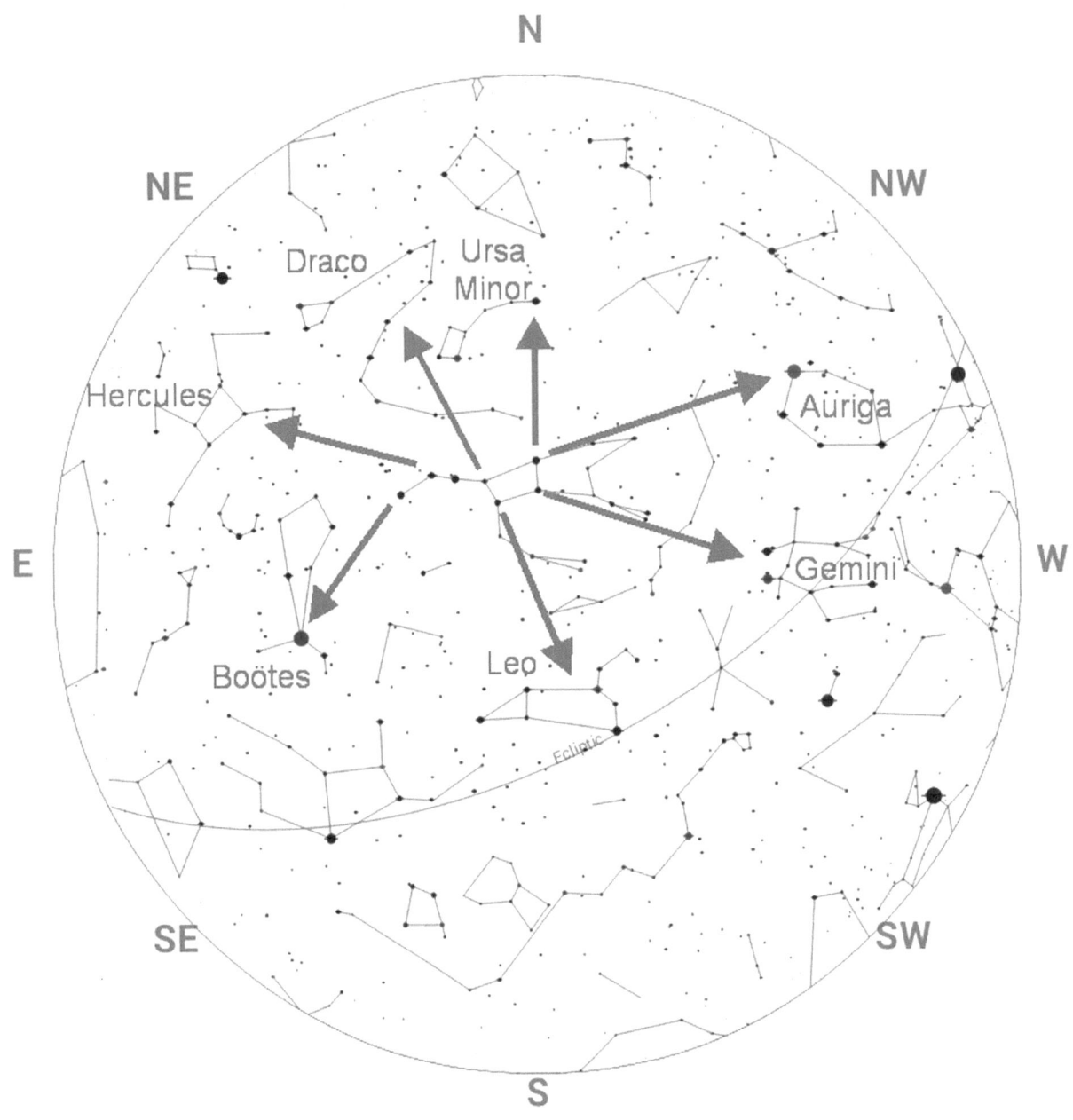

This chart can help you to use Ursa Major to identify other constellations. It depicts the night sky at the following times:

Early April	Late April	Early May	Late May
11 p.m.	10 p.m.	9 p.m.	8 p.m.

The Heart of the
Milky Way

This chart can help you to find the center of the Milk Way, Sagittarius and the other summer constellations. It depicts the night sky at the following times:

Late July	Early August	Late August	Early September
11 p.m.	10 p.m.	9 p.m.	8 p.m.

The Andromeda
Galaxy

This chart can help you to find the Andromeda Galaxy and the autumn constellations. It depicts the night sky at the following times:

Late October	Early November	Late November	Early December
11 p.m.	10 p.m.	9 p.m.	8 p.m.